1-Hour
Geometry Review Guide
For the End-of-Course,
SAT, ACT, and ASSET tests

1-Hour Geometry Review Guide For the End-of-Course, SAT, ACT, and ASSET tests

Everything you need to know, want to know, or just plain forgot!

Brenda Voyles

authorHOUSE®

AuthorHouse™
1663 Liberty Drive
Bloomington, IN 47403
www.authorhouse.com
Phone: 1-800-839-8640

First published by AuthorHouse   07/13/2011

ISBN: 978-1-4634-3148-8 (sc)
ISBN: 978-1-4634-3147-1 (ebk)

Library of Congress Control Number: 2011911931
Printed in the United States of America

Begin With Basics

Point – A point in space that has no dimension
       Represented by a dot and named by a capital letter

A
•
Point A

Line – Determined by 2 points and has one dimension named
      by any 2 points on that line

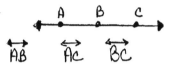

Line segment –Contains 2 endpoints and all the points between them
             Named by the 2 endpoints

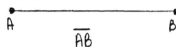

Ray – Consists of an endpoint and all points that lie on the ray
     side of endpoints

Angle – Consists of two different rays with a common endpoint

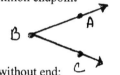

∢ B
∢ ABC
∢ CBA

Plane – A flat 2 dimensional shape that extends without end;
       determined by 3 points

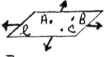

Plane ℓ

Numbers and measurements are equal

m∢A = m∢B   AB = CD
  90 = 90   4 = 4

Figures, shapes, angles, arcs, and segments are congruent – same
shape and different size

∢A ≅ ∢B $\overline{AB} ≅ \overline{CD}$

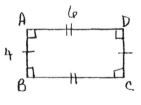

Figures, shapes, and segments are similar – same shape, different size

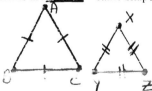

△ABC ~ △XYZ
Angles are always
Congruent
(same measure)

1

<u>Collinear Points</u> – points all on one line

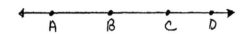

<u>Coplanar Points</u> – points all in one plane

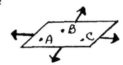

<u>Adjacent Angles</u> – Angles that share a common vertex and side, but have no common
  interior points

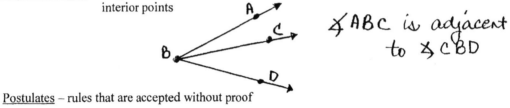

∡ABC is adjacent
  to ∡CBD

<u>Postulates</u> – rules that are accepted without proof

<u>Theorems</u> – rules that are proved

Properties

Addition/Subrtaction Property

if   a = b
then   a+c = b+c
           or
          a-c = b-c
If two equations or
numbers are equal,
they will remain equal
if you add or subtract
the same number to both
sides

Multiplication/Division Property

if   a = b
then   a x c = b x c
           or
          a/c = b/c
If two equations or numbers
are equal, they will remain
equal if you multiply or
divide by the same number
on both sides

Substitution Property

if   a = b
then   "a" can be substituted
for "b" in any equation
or expression

Reflexive Property of Equality

for numbers   A = A   5 = 5

for segments $\overline{AB} = \overline{AB}$

for angles   m $\angle$ A = m $\angle$ A

Symmetric Property of Equality   for numbers   if a = b, then b = a

for segments   if $\overline{AB} = \overline{CD}$, then $\overline{CD} = \overline{AB}$

for angles   if m$\angle$A = m$\angle$B, then m$\angle$B = m$\angle$A

<u>Transitive Property of Equality</u>   for numbers   if $a = b$ and $b = c$, then $a = c$

for segments if $\overline{AB} = \overline{CD}$ and $\overline{CD} = \overline{EF}$,
then $\overline{AB} = \overline{EF}$

for angles   if $m\angle A = m\angle B$ and $m\angle B = m\angle C$,
then $m\angle A = m\angle C$

Simple Proof Using Properties

AB = CD

Show AC = BD

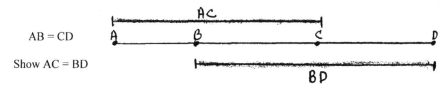

Equation	Explanation	Reason
AB = CD	Marked on diagram	Given
AC = AB + BC	Add lengths of adjacent segments	Segment Addition Postulate
BD = BC + CD	Add lengths of adjacent segments	Segment Addition Postulate
AB + BC = CD + BC	Add BC to both sides	Addition Property of Equality
AC = BD	Substitute AC for AB + BC And BC for BC + CD	Substitution Property of Equality

Conditional Statements

<u>Conditional statements</u> – are if – then statements

If........................, then................... .
(hypothesis)   (conclusion)

Conditional statements can be true or false:

      True – must prove the conclusion is true every time
          the hypothesis is true
      False – give only one counterexample

<u>Counterexample</u> – a specific example that shows the statement (conjecture) is false

<u>Conjecture</u> – an unproven statement that is based on observation

      Example: Conjecture – All prime numbers are odd
                Counterexample – 2 is a prime number that is not odd

<u>Converse</u> – is formed by interchanging the hypothesis and conclusion

<u>Negation</u>—the opposite of the original statement

<u>Inverse</u> – negate the hypothesis and conclusion in conditional statement

<u>Counterpositive</u> – negate the converse

Example: Conditional – If m∠A = 90, then ∠A is a right angle
          Converse – If ∠A is a right angle, then m∠A = 90
          Inverse – If m∠A = 90, then ∠A is not a right angle
          Counterpositive – If ∠A is not a right angle, then m∠A = 90

A conditional statement and its' contrapositive are either **<u>both</u>** true or **<u>both</u>** false

Similarly, the converse and inverse of a conditional statement are either **<u>both</u>** true or **<u>both</u>** false.

<u>Biconditional statement</u> – a statement that contains the phrase "if and only if" and is written if a conditional and its converse are both true

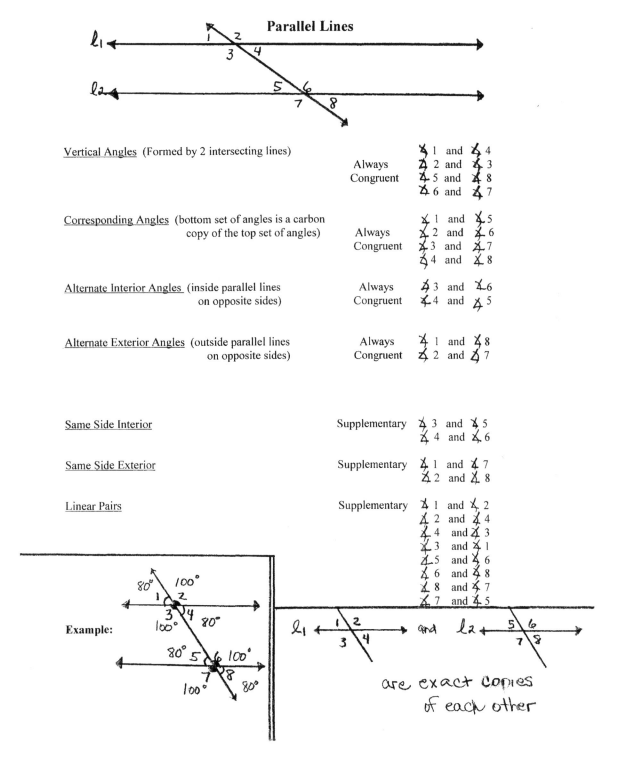

Parallel Lines

Vertical Angles (Formed by 2 intersecting lines)

Always Congruent

∢ 1 and ∢ 4
∢ 2 and ∢ 3
∢ 5 and ∢ 8
∢ 6 and ∢ 7

Corresponding Angles (bottom set of angles is a carbon copy of the top set of angles)

Always Congruent

∢ 1 and ∢ 5
∢ 2 and ∢ 6
∢ 3 and ∢ 7
∢ 4 and ∢ 8

Alternate Interior Angles (inside parallel lines on opposite sides)

Always Congruent

∢ 3 and ∢ 6
∢ 4 and ∢ 5

Alternate Exterior Angles (outside parallel lines on opposite sides)

Always Congruent

∢ 1 and ∢ 8
∢ 2 and ∢ 7

Same Side Interior

Supplementary

∢ 3 and ∢ 5
∢ 4 and ∢ 6

Same Side Exterior

Supplementary

∢ 1 and ∢ 7
∢ 2 and ∢ 8

Linear Pairs

Supplementary

∢ 1 and ∢ 2
∢ 2 and ∢ 4
∢ 4 and ∢ 3
∢ 3 and ∢ 1
∢ 5 and ∢ 6
∢ 6 and ∢ 8
∢ 8 and ∢ 7
∢ 7 and ∢ 5

Example:

are exact copies of each other

Polynomial Classifications

3-sided   Triangle   tri – 3

4-sided   Quadrilateral

5-sided   Pentagon   (think of the Pentagon in Washington)

6-sided   Hexagon   (666—bad luck; put a "hex" on someone)

7-sided   Heptagon   (sounds like 'heaven' and 'seven')

8-sided   Octagon   (octopus has 8 legs)

9-sided   Nonagon   ('nueve' in Spanish means nine)

10-sided   Decagon   (think of 10 years – a decade)

11-sided   Undegon   (go 'under' the chicken to get the 12th egg)

12-sided   Dodecagon   (think of 12 – a dozen)

Degrees in a Polygon

*You must first know that all triangles consist of 180 degrees.

$180°$

A

B   C

All 4-sided figures have 360

A   D

$360°$

B   C

4 – sided quadrilateral

To find – pick a vertex
-- connect to all other vertices
-- 2 triangles formed
     $(180 + 180 = 360)$

A   D

$180°$

$180°$

B   C

To find – pick a vertex
-- connect to all other vertices
-- 3 triangles formed
     $(180 + 180 + 180 = 540)$

5 – sided pentagon

A

B $180°$ $180°$ E

$180°$

C   D

To find – pick a vertex
        -- connect to all other vertices
        -- 4 triangles formed
        (180 + 180 + 180 + 180 = <u>720</u>)

6 – sided hexagon

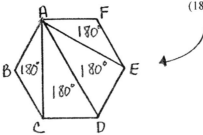

SHORTCUT - for Finding Degrees in a Polynomial

$$(N - 2) \times 180$$
       ↑
   number of sides

Example: Octagon – 8 sides $(8 - 2) \times 180$
                                6   $\times$ 180
                                    ↓
                                <u>1080</u>

General Rule

3 sides – 1 triangle = 180 degrees
4 sides – 2 triangles = 360 degrees
5 sides – 3 triangles = 540 degrees
6 sides – 4 triangles = 720 degrees
N sides – (N – 2) triangles = (N – 2) x 180

To Find Interior and Exterior angles of a Regular Polygon

Regular Polygon – All sides and all angles congruent

Interior

*To find interior angles of a regular polygon, find total degrees and divide by number of angles

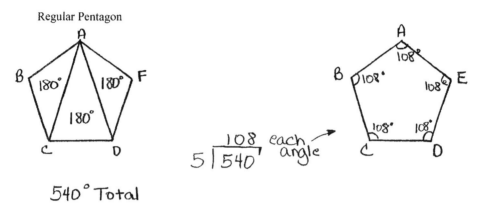

Regular Pentagon

540° Total

$5\overline{)540}$ = 108 each angle

Exterior

*All exterior angles in a regular polygon are congruent
  *Exterior and interior angles are linear pairs

Linear Pairs – 2 angles that form a straight line or add up to 180 degrees

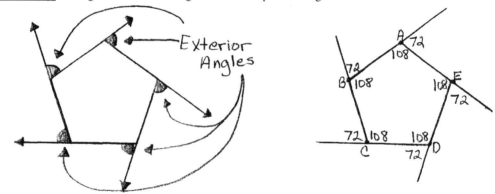

Exterior Angles

SHORTCUT – Exterior angles <u>always</u> add up to 360—so, divide 360 by number of total angles

$5\overline{)360}$ = 72

each exterior angle = 72°

Quadrilaterals

Square *Opposite sides are parallel

  *All sides are congruent

  *All angles are right angles (90degrees)

  *Both diagonals are congruent and bisect

  *Diagonals are perpendicular (form right angles)

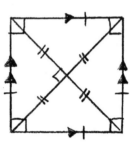

Rhombus *Opposite sides are parallel

  *All sides are congruent

  *Opposite angles are congruent

  *Consecutive angles are supplementary (add up to 180 degrees)

  *Diagonals bisects

  *Diagonals are perpendicular (form right angles)

Consecutive angles = 180°

<u>Rectangles</u>   *Opposite sides are parallel

*Opposite sides are congruent

*All angles are right angles (90 degrees)

*Diagonals congruent and bisect

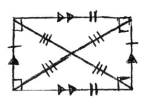

<u>Parallelograms</u>   *Opposite sides are parallel

*Opposite sides are congruent

*Opposite angles are congruent

*Consecutive angles are supplementary (add up to 180 degrees)

*Diagonals bisect

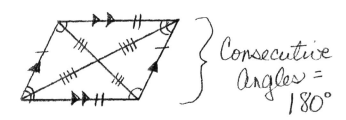

Trapezoids – Only 2 sides are parallel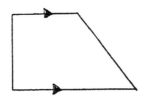

Isosceles Trapezoid   *only 2 sides are parallel ⟶

*base angles are congruent

*top angles are congruent

*diagonals are congruent

*consecutive side angles are supplementary (add up to 180 degrees)

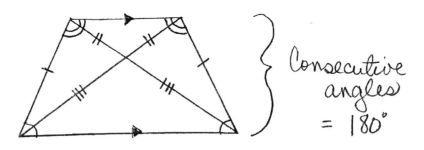

Consecutive angles = 180°

Kite   *has two pairs of consecutive congruent sides

*opposite sides are not congruent

*Diagonals are perpendicular

*Exactly one pair of opposite angles are congruent

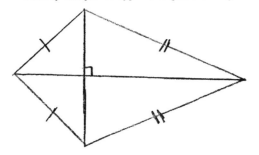   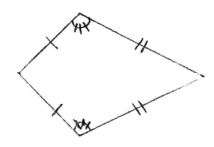

Proving a Quadrilateral is a Parallelogram

*Show ABCD (Both pairs of opposite sides are congruent)

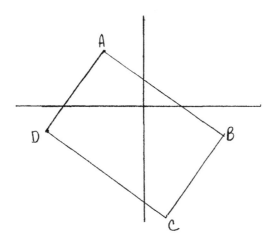

- Create 2 right triangles from opposite
  Sides of ABCD
- Since legs $XA = 2$ $YC = 2$
  then
  $$\overline{XA} \cong \overline{YC}$$
- Since legs $XD = 4$ $YB = 4$
  then
  $$\overline{XD} \cong \overline{YB}$$
- Since
  $$m \angle x = 90 \quad m \angle y = 90$$
  then
  $$\angle x \cong \angle y$$

So $\triangle XAD \cong \triangle YCB$
  (Side-Angle-Side)
  SAS

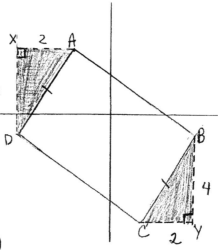

15

- Create 2 right triangles
  From opposite sides
  of $\overline{ABCD}$

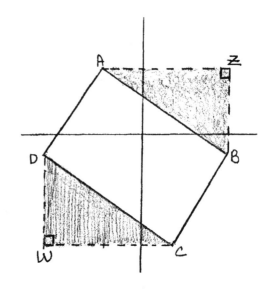

- Since legs $ZA = 6$ $WC = 6$
  then
  $\overline{ZA} \cong \overline{WC}$

- Since legs $ZB = 4$ $WD = 4$
  then
  $\overline{ZB} \cong \overline{WD}$

- Since $m\angle Z = 90$ $m\angle W = 90$
  then
  $\angle Z \cong \angle W$

So $\triangle ZAB \cong \triangle WCD$
  (Side - Angle - Side)
  SAS

* Because opposite sides
  $\overline{AD}$ and $\overline{CB}$
  $\overline{AB}$ and $\overline{CD}$
  are congruent
Then ABCD is a Parallelogram

Classifying Triangles

Classifying triangles by sides   Scalene – no congruent sides

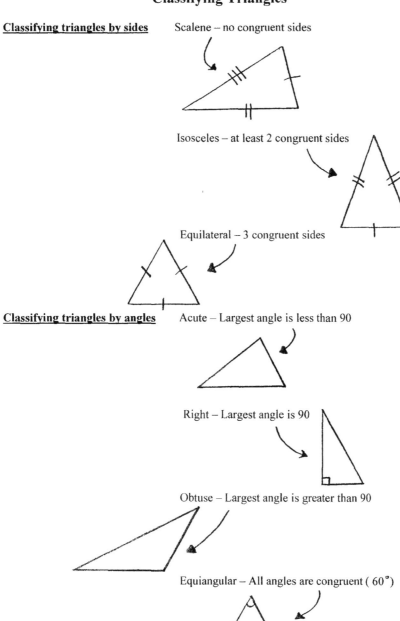

Isosceles – at least 2 congruent sides

Equilateral – 3 congruent sides

Classifying triangles by angles   Acute – Largest angle is less than 90

Right – Largest angle is 90

Obtuse – Largest angle is greater than 90

Equiangular – All angles are congruent (60°)

Interior Angles   The sum of all interior angles is always = 180 degrees

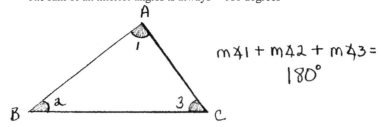

$$m\angle 1 + m\angle 2 + m\angle 3 = 180°$$

Exterior Angles   The measure of an exterior angle of a triangle is equal to the sum of the measures of the two nonadjacent interior angles

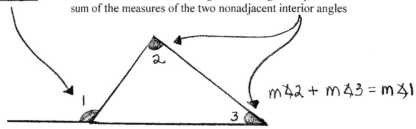

$$m\angle 2 + m\angle 3 = m\angle 1$$

Supplementary Angles   Two angles whose measures have a sum of 180 degrees

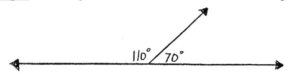

Complementary Angles   Two angles whose measures have a sum of 90 degrees

Adjacent Angles   Two angles that share a common vertex and side, but have no common interior points

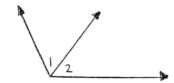

Linear Pair   Two adjacent angles whose noncommon sides are opposite rays

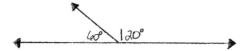

Congruence

Two geometric figures are **congruent** if they have exactly the same size and shape

*If 2 figures are congruent ($\cong$) all corresponding angles and all corresponding sides are congruent

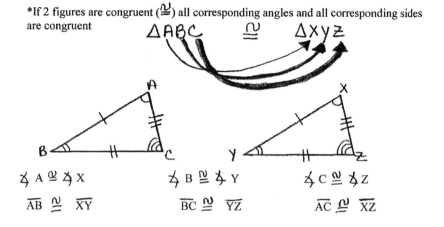

$\angle A \cong \angle X$ $\angle B \cong \angle Y$ $\angle C \cong \angle Z$

$\overline{AB} \cong \overline{XY}$ $\overline{BC} \cong \overline{YZ}$ $\overline{AC} \cong \overline{XZ}$

Proving Congruence

S – S – S   side – side – side

S – A – S   side – angle – side

HL   hypotenuse – leg

A – S – A   angle – side – angle

A – A – S   angle – angle – side

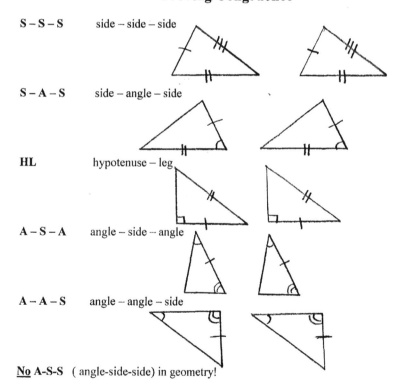

No A-S-S (angle-side-side) in geometry!

*If two sides of a triangle are congruent, then the angles opposite them are congruent (base angles)

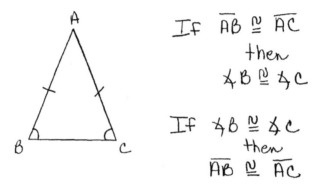

If $\overline{AB} \cong \overline{AC}$
then
$\angle B \cong \angle C$

If $\angle B \cong \angle C$
then
$\overline{AB} \cong \overline{AC}$

*If two base angles of a triangle are congruent, then the sides opposite them are congruent

---

*If a triangle is equilateral, then it is equiangular

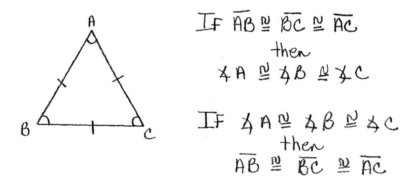

If $\overline{AB} \cong \overline{BC} \cong \overline{AC}$
then
$\angle A \cong \angle B \cong \angle C$

If $\angle A \cong \angle B \cong \angle C$
then
$\overline{AB} \cong \overline{BC} \cong \overline{AC}$

*If a triangle is equiangular, then it is equilateral

Similar

Similar – same shape, different size; always proportional

EXAMPLE: How to find a missing length

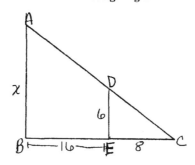

1. Draw two similar drawings separately next to each other

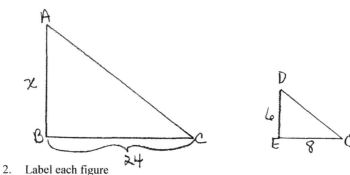

2. Label each figure
3. Find something you know about both figures and drop it down or on the same level of proportion and label

$$\frac{x}{24} \quad \overset{Height}{\underset{Base}{=}} \quad \frac{6}{8}$$

4. Find what you **need** to know in figure (x), match it to other figure, and drop down on same level and label
5. Cross multiply values you know and divide values that has an unknown

$$\frac{x}{24} = \frac{6}{8}$$

$$24 \cdot 6 = 8x$$
$$144 = 8x$$
$$\frac{144}{8} = \frac{8x}{8}$$
$$\boxed{10 = x}$$

*Remember lengths in similar figures are proportional, **but all angles are congruent.**

EXAMPLE:

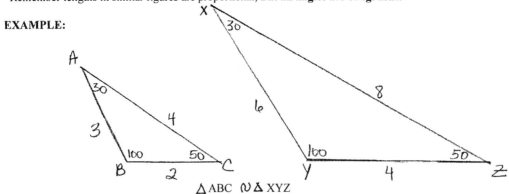

$\triangle$ ABC $\sim$ $\triangle$ XYZ

means

$\overline{AB} \sim \overline{XY}$ $\overline{BC} \sim \overline{YZ}$ $\overline{AC} \sim \overline{XZ}$ $\Big\}$ All lengths

$\dfrac{3}{6} = \dfrac{1}{2}$ $\dfrac{2}{4} = \dfrac{1}{2}$ $\dfrac{4}{8} = \dfrac{1}{2}$ $\Big\}$ are proportional

$\angle A \sim \angle X$ $\angle B \sim \angle Y$ $\angle C \sim \angle Z$ $\Big\}$ All angles

$30 = 30$ $100 = 100$ $50 = 50$ $\Big\}$ are congruent

---

*Anytime you have **1** figure that is a right triangle – it is a Pythagorean theorem problem

If
1 Figure
Given

15

c

20

You can always
assume anything
coming out of flat
ground is perpendicular (1)
– forms a right angle

15 foot Flagpole
20 foot shadow
– how long is rope if
attached to top of
Flagpole and end
of shadow?

$a^2 + b^2 = c^2$
$15^2 + 20^2 = c^2$
$225 + 400 = c^2$
$625 = c^2$
$\sqrt{625} = \sqrt{c^2}$
$25 = c$
$\boxed{25 \text{ foot rope}}$

*Anytime you have **2** similar figures (same shape – different size or all angles equal)

The problem represents figures that are similar, which are **set up as a proportion**

EXAMPLE:

If
2 figures
Given

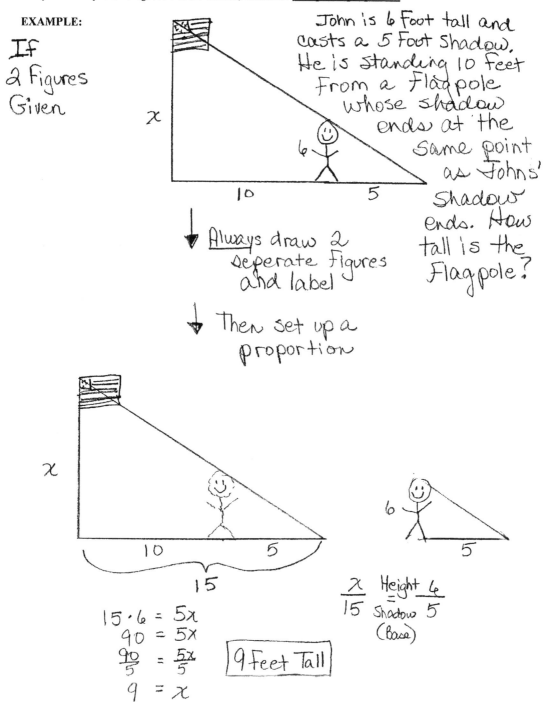

John is 6 foot tall and casts a 5 foot shadow. He is standing 10 feet from a flagpole whose shadow ends at the same point as Johns' shadow ends. How tall is the flagpole?

Always draw 2 seperate figures and label

Then set up a proportion

$15 \cdot 6 = 5x$

$90 = 5x$

$\frac{90}{5} = \frac{5x}{5}$

$9 = x$

$\frac{x}{15} \begin{array}{l} \text{Height} \\ = \\ \text{Shadow} \end{array} \frac{6}{5}$
(Base)

9 Feet Tall

Proving Similarity

Side – Side – Side (SSS) Similarity Theorem – if the corresponding side lengths of two triangles are proportional, then the triangles are similar

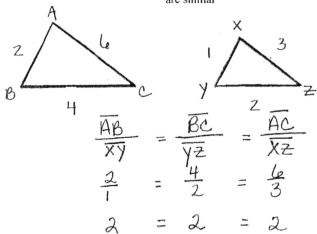

$$\frac{\overline{AB}}{\overline{XY}} = \frac{\overline{BC}}{\overline{YZ}} = \frac{\overline{AC}}{\overline{XZ}}$$

$$\frac{2}{1} = \frac{4}{2} = \frac{6}{3}$$

$$2 = 2 = 2$$

Angle – Angle (AA) Similarity Postulate – If two angles of one triangle are congruent to two angles of another triangle, then the two triangles are similar

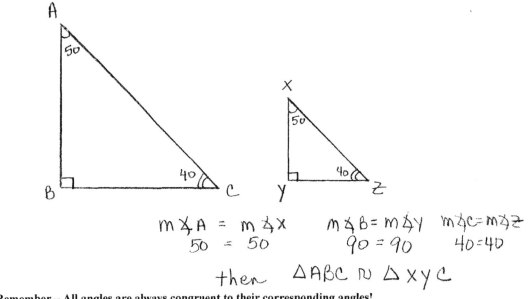

$$m \angle A = m \angle X \qquad m \angle B = m \angle Y \qquad m \angle C = m \angle Z$$
$$50 = 50 \qquad\qquad 90 = 90 \qquad\qquad 40 = 40$$

$$\text{then} \quad \triangle ABC \sim \triangle XYC$$

*Remember – <u>All</u> angles are <u>always</u> congruent to their corresponding angles!

Scale Factor

*If 2 figures are similar (same shape, different size)

Determine if the **scale drawing** – drawing that is the same shape as the object it represents is getting either larger or smaller – **a dilation**

Check to see when you go from A ------ A′

↑   ↑

Original   new figure

If it is getting larger – scale factor will be > 1

If it is getting smaller – scale factor will be 0 < S < 1 (fraction smaller than 1)

To Find Scale Factor

Fraction bar → _____ length of new figure ⎰ measurement
                 length of original figure ⎱ of corresponding lengths

EXAMPLE:

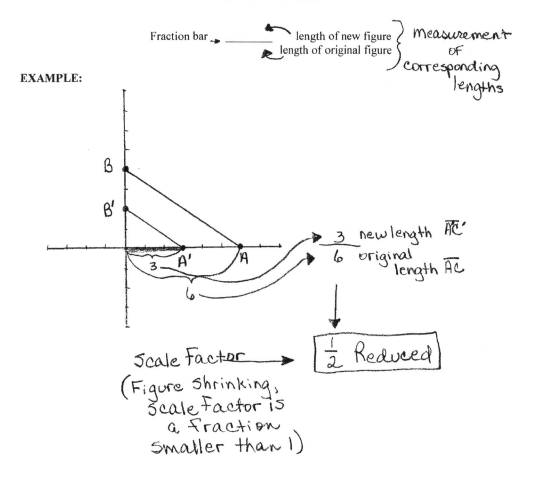

$\dfrac{3}{6}$ new length $\overline{AC'}$ / original length $\overline{AC}$

$\dfrac{1}{2}$ Reduced

Scale Factor → (Figure shrinking, scale factor is a fraction smaller than 1)

25

A Similarity Transformation

Dilation – transformation that stretches or shrinks a figure to create a similar figure

*dilation, a figure is enlarged or reduced with respect to a fixed point (center or dilation) which is often the origin (0, 0)

As previously discussed, a scale factor controls whether the figure is a reduction (scale factor $0 < S < 1$) or an enlargement (scale factor $S > 1$)

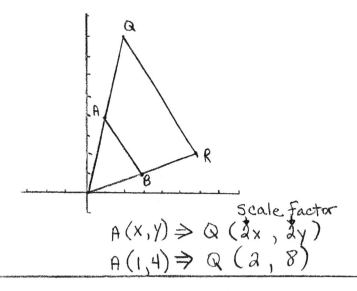

scale factor

$A(x, y) \Rightarrow Q(2x, 2y)$

$A(1, 4) \Rightarrow Q(2, 8)$

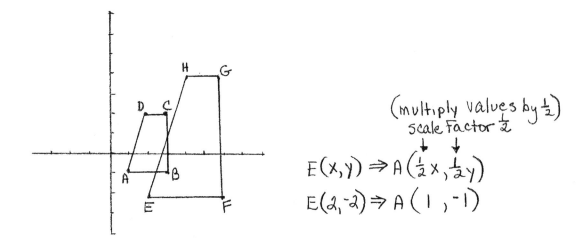

(multiply values by $\frac{1}{2}$)

scale factor $\frac{1}{2}$

$E(x, y) \Rightarrow A(\frac{1}{2}x, \frac{1}{2}y)$

$E(2, -2) \Rightarrow A(1, -1)$

Geometric Mean

$$\frac{a}{x} = \frac{x}{b}$$

$$\frac{a}{x} = \frac{x}{b} \longrightarrow \begin{aligned} x \cdot x &= a \cdot b \\ x^2 &= ab \\ x &= \sqrt{ab} \end{aligned}$$

EXAMPLE:

Find Geometric Mean 12 and 20

$$\frac{12}{a} = \frac{a}{20}$$

$$a \cdot a = 12 \cdot 20$$

$$a^2 = 240$$

To simplify

$$\sqrt{a^2} = \sqrt{240}$$

$$\begin{array}{r|l} 2 & 240 \\ 2 & 120 \\ 2 & 60 \\ 2 & 30 \\ 3 & 15 \\ & 5 \end{array}$$

$$a = \sqrt{2 \cdot 2 \cdot 2 \cdot 2 \cdot 3 \cdot 5}$$

$$\sqrt{4} \quad \sqrt{4}$$

$$a = 2 \cdot 2\sqrt{15}$$

$$a = 4\sqrt{15}$$

or

$$a \approx 15.5$$

Midsegment

*of a triangle

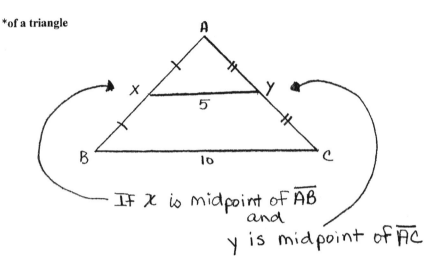

IF x is midpoint of $\overline{AB}$
and
y is midpoint of $\overline{AC}$

Then $\overline{xy}$ is midsegment of $\triangle ABC$ and

$$\overline{xy} = \tfrac{1}{2}\overline{BC}$$
$$5 = \tfrac{1}{2}(10)$$
$$5 = 5$$

*of a trapezoid

IF x is midpoint of $\overline{AD}$
and
y is midsegment of $\overline{BC}$

Then $\overline{xy}$ is midsegment of $ABCD$

$$xy = \tfrac{1}{2}(\overline{AB} + \overline{CD})$$
$$15 = \tfrac{1}{2}(30)$$
$$15 = 15$$

EXAMPLE:

If you **don't** know the median,
add the 2 bases and divide the sum
by 2 (take ½)

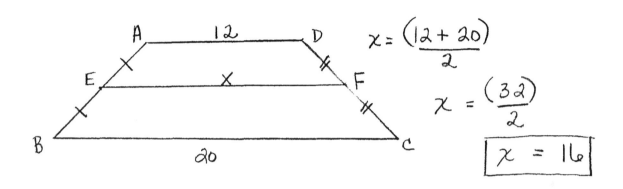

$$x = \frac{(12 + 20)}{2}$$

$$x = \frac{(32)}{2}$$

$$\boxed{x = 16}$$

EXAMPLE:

If you know median, double median
and set equal to sum of 2 bases

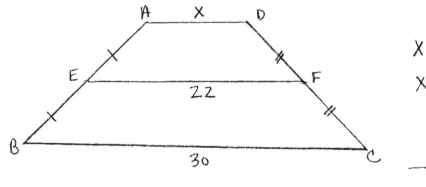

$$x + 30 = 22 \cdot 2$$

$$x + 30 = 44$$

$$x + 30 = 44$$
$$ -30 \quad -30$$

$$\boxed{x = 14}$$

Perpendicular Bisectors

*To bisect a segment means to intersect a segment at its midpoint

   *If this intersection is perpendicular to the segment, then it is called a **perpendicular bisector**

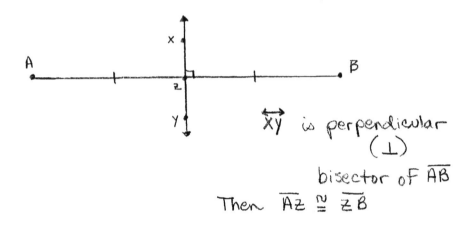

$\overleftrightarrow{XY}$ is perpendicular ($\perp$) bisector of $\overline{AB}$

Then $\overline{AZ} \cong \overline{ZB}$

Concurrency of Perpendicular Bisectors of a Triangle

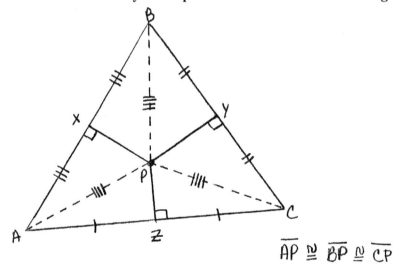

$\overline{AP} \cong \overline{BP} \cong \overline{CP}$

*The perpendicular bisectors of a triangle intersect at a point that is equildistance from the vertices of the triangle

Angle Bisectors

*An **Angle Bisector** is a ray that divides an angle into two congruent adjacent angles

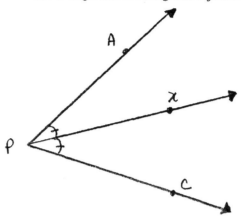

Ray $\overrightarrow{PX}$ bisects $\angle APC$, $\angle APX \cong \angle CPX$

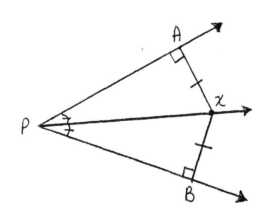

Ray $\overrightarrow{PX}$ bisects $\angle APB$
If $\overline{AX} \perp \overrightarrow{PA}$ and $\overline{BX} \perp \overrightarrow{PB}$
Then $\overline{AX} \cong \overline{BX}$

Inequalities in Triangles

*If one side of a triangle is longer than another side, then the angle opposite the longest side is larger than the angle opposite the shorter side

EXAMPLE:

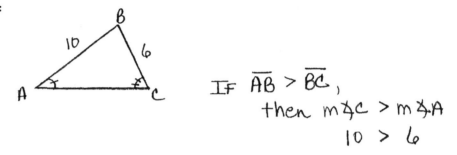

IF $\overline{AB} > \overline{BC}$,
then $m \angle C > m \angle A$
$10 > 6$

*If one angle of a triangle is larger than another angle, then the side opposite the larger angle is longer than the side opposite the smaller angle

EXAMPLE:

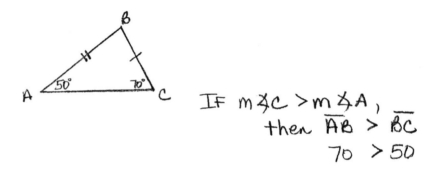

IF $m \angle C > m \angle A$,
then $\overline{AB} > \overline{BC}$
$70 > 50$

*The longest side of any triangle is longer than either of the other two sides, but it is less than the sum of the smaller two lengths

EXAMPLE:

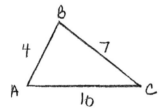

$\overline{AC}$ is longest side
$\overline{AC} > \overline{AB}$ and
$\overline{AC} > \overline{BC}$
but
$\overline{AC} < \overline{AB} + \overline{BC}$

Classifying Triangles by Angles
Using Side Lengths

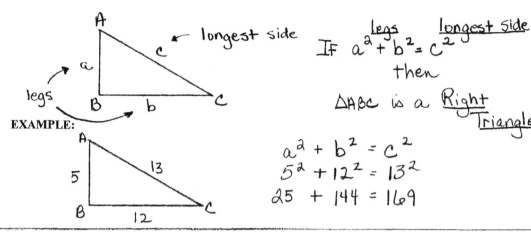

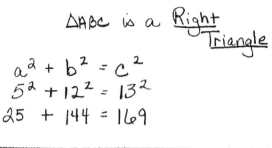

$$\text{If } a^2 + b^2 = c^2$$
$$\text{then}$$

$\triangle ABC$ is a $\underline{\text{Right}}$ $\underline{\text{Triangle}}$

$$a^2 + b^2 = c^2$$
$$5^2 + 12^2 = 13^2$$
$$25 + 144 = 169$$

longest side

EXAMPLE:

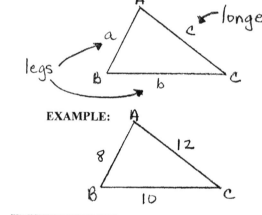

longest side

$$\text{If } a^2 + b^2 > c^2$$
$$\text{then}$$

$\triangle ABC$ is an $\underline{\text{Acute}}$ $\underline{\text{Triangle}}$

$$a^2 + b^2 > c^2$$
$$8^2 + 10^2 > 12^2$$
$$64 + 100 > 144$$
$$164 \quad > 144$$

EXAMPLE:

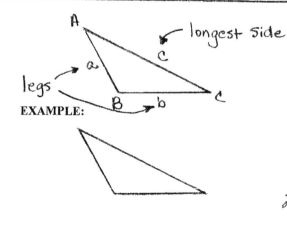

legs

longest side

EXAMPLE:

$$\text{If } a^2 + b^2 < c^2$$
$$\text{then}$$

$\triangle ABC$ is an $\underline{\text{Obtuse}}$ $\underline{\text{Triangle}}$

$$a^2 + b^2 < c^2$$
$$5^2 + 6^2 < 10^2$$
$$25 + 36 < 100$$
$$61 \quad < 100$$

$\triangle ABC$ is an $\underline{\text{Obtuse}}$ $\underline{\text{Triangle}}$

Pythagorean Theorem

*Use Pythagorean theorem to find the missing side of a right triangle if given 2 sides

$$a^2 + b^2 = c^2$$

shortest side (leg)   middle side (leg)   longest side (hypotenuse)

EXAMPLE: If C is unknown

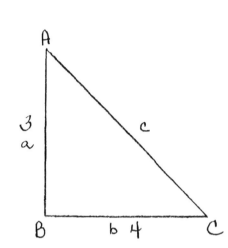

$$a^2 + b^2 = c^2$$
$$3^2 + 4^2 = c^2$$
$$9 + 16 = c^2$$
$$25 = c^2$$
$$\sqrt{25} = \sqrt{c^2}$$
$$\boxed{5 = C}$$

to get rid of an exponent, take $\sqrt{c^2} = c$

* If c is unknown, always __add__ $a^2 + b^2$

EXAMPLE: If a or b is unknown

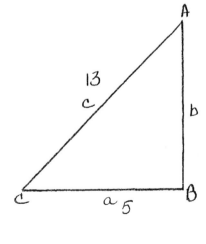

$$a^2 + b^2 = c^2$$
$$5^2 + b^2 = 13^2$$
$$25 + b^2 = 169$$
$$-25 \qquad -25$$
$$b^2 = 144$$
$$\sqrt{b^2} = \sqrt{144}$$
$$\boxed{b = 12}$$

to get rid of an exponent, take $\sqrt{b^2} = b$

* If you know C and either a or b, Always __subtract__

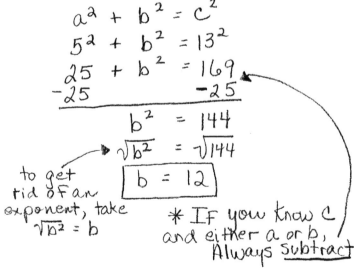

Special Right Triangles

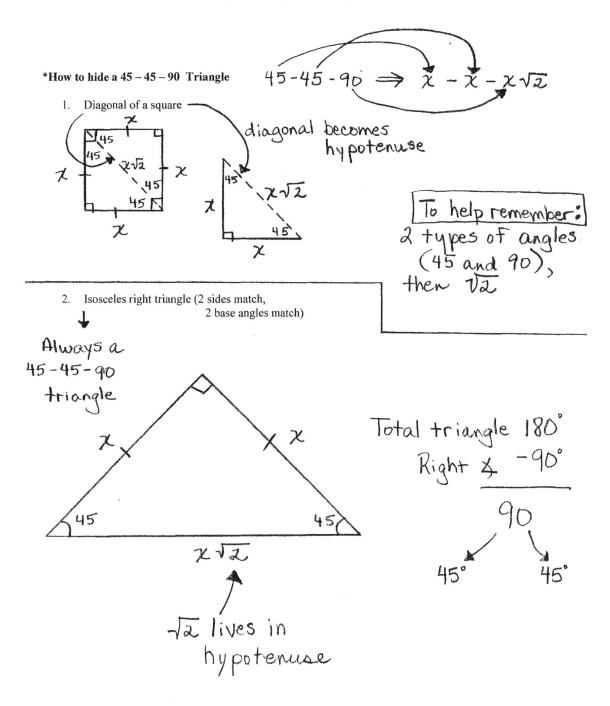

*How to hide a 45 – 45 – 90 Triangle

$$45 - 45 - 90 \Rightarrow x - x - x\sqrt{2}$$

1. Diagonal of a square

diagonal becomes hypotenuse

To help remember:
2 types of angles
(45 and 90),
then $\sqrt{2}$

2. Isosceles right triangle (2 sides match,
2 base angles match)

Always a
45 – 45 – 90
triangle

Total triangle $180°$
Right $\angle$ $-90°$
_____
90

$45°$ $45°$

$\sqrt{2}$ lives in hypotenuse

EXAMPLE: If $\sqrt{2}$ is in the hypotenuse

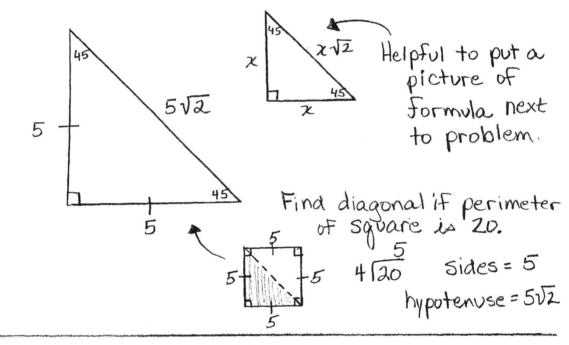

Helpful to put a
picture of
formula next
to problem.

Find diagonal if perimeter
of square is 20.

$$4\overline{)20} \quad \overset{5}{} \quad \text{sides} = 5$$

hypotenuse $= 5\sqrt{2}$

Example: If $\sqrt{2}$ is not in the hypotenuse

Find length of sides of a square
if the diagonal is 10.

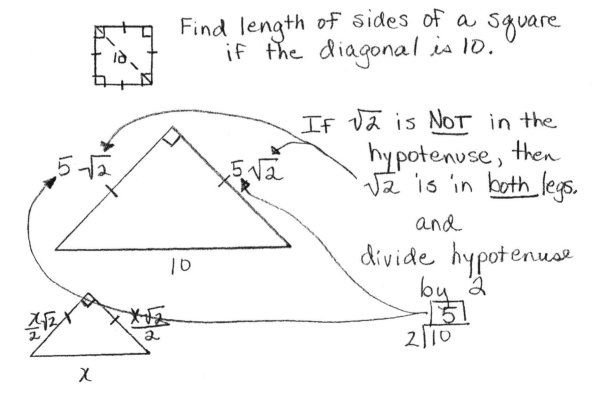

If $\sqrt{2}$ is <u>NOT</u> in the
hypotenuse, then
$\sqrt{2}$ is in <u>both legs</u>.

and
divide hypotenuse
by 2

$$2\overline{)10} \quad \boxed{5}$$

36

*How to hide a 30 – 60 – 90 Triangle

$30-60-90 \Rightarrow x - x\sqrt{3} - 2x$

1. Equilateral triangle (All sides congruent and angles = 60 degrees)

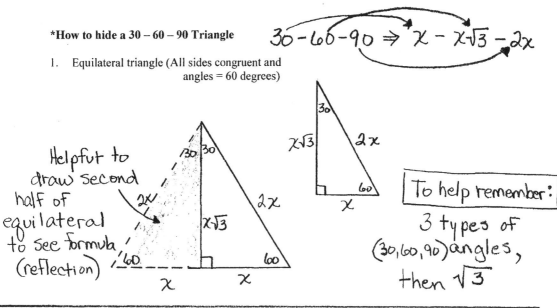

Helpful to draw second half of equilateral to see formula (reflection)

To help remember:
3 types of $(30,60,90)$ angles, then $\sqrt{3}$

2. Apothem of a hexagon

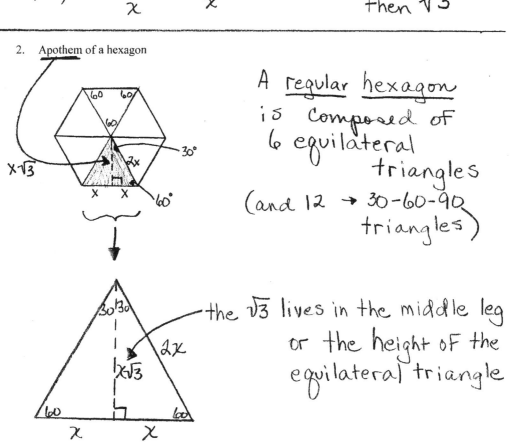

A regular hexagon is composed of 6 equilateral triangles (and 12 → 30-60-90 triangles)

the $\sqrt{3}$ lives in the middle leg or the height of the equilateral triangle

37

EXAMPLE: If $\sqrt{3}$ is in the middle leg (height of equilateral triangle)

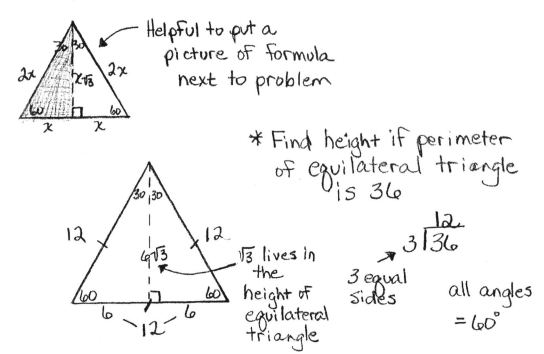

Helpful to put a picture of formula next to problem

* Find height if perimeter of equilateral triangle is 36

$\sqrt{3}$ lives in the height of equilateral triangle

3 equal sides

all angles = 60°

$3\overline{)36}$ → $\frac{12}{}$

EXAMPLE: If $\sqrt{3}$ is not in the middle leg (height of equilateral triangle)

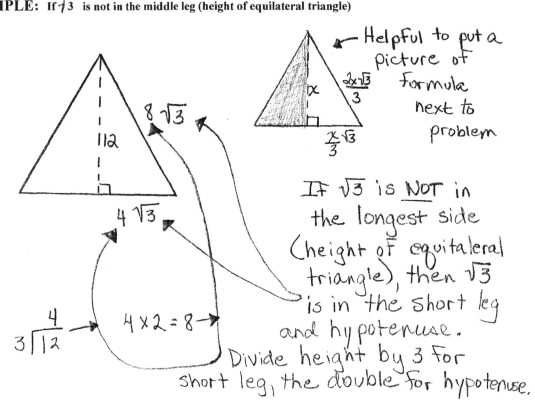

Helpful to put a picture of formula next to problem

If $\sqrt{3}$ is NOT in the longest side (height of equilateral triangle), then $\sqrt{3}$ is in the short leg and hypotenuse. Divide height by 3 for short leg, the double for hypotenuse.

$3\overline{)12}$ → $\frac{4}{}$ $4 \times 2 = 8$ →

38

Similar Right Triangles

*When the altitude is drawn to the hypotenuse of a right triangle, the two newly formed smaller
  Triangles are similar to the original triangle and to each other

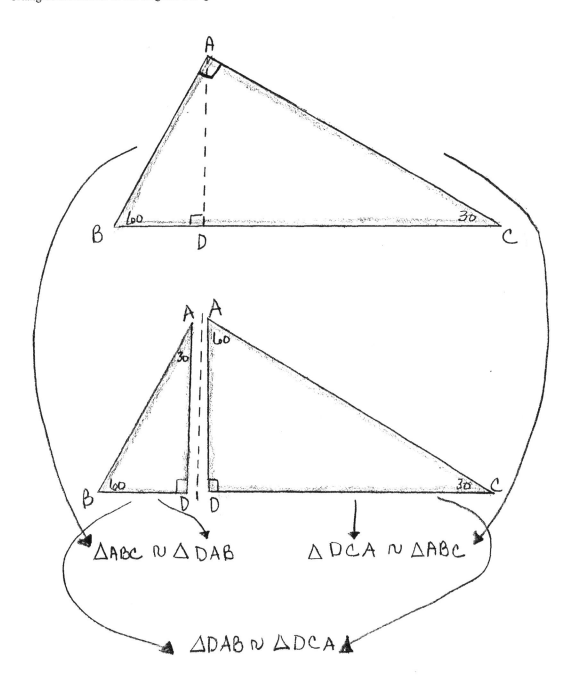

$\triangle ABC \sim \triangle DAB$

$\triangle DCA \sim \triangle ABC$

$\triangle DAB \sim \triangle DCA$

Using Trigonometric Ratios
To Solve Right Triangles

Tangent, Sine, Cosine ratios can be used to find
unknown lengths and angle measures of right triangles

$$\text{Sin } A \quad \frac{\text{opposite}}{\text{hypotenuse}} \quad \frac{BC}{AB} \quad\Big|\Big|\quad \text{Sin}^{-1} \frac{BC}{AB} = m \angle A$$

$$\text{Cos } A \quad \frac{\text{adjacent}}{\text{hypotenuse}} \quad \frac{AC}{AB} \quad\Big|\Big|\quad \text{Cos}^{-1} \frac{AC}{AB} = m \angle A$$

$$\text{Tan } A \quad \frac{\text{opposite}}{\text{adjacent}} \quad \frac{BC}{AC} \quad\Big|\Big|\quad \text{Tan}^{-1} \frac{BC}{AC} = m \angle A$$

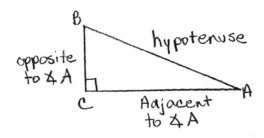

EXAMPLE:

$$\text{Tan } A = \frac{\text{opposite}}{\text{adjacent}} = \frac{18}{12} = \frac{3}{2}$$

$$= 1.5$$

$$\text{Tan}^{-1} 1.5 = m \angle A$$

$$\approx 56.3099324...$$

$$m \angle A \approx 56.31$$

Transformations

Translation – "Slide" x right or left , then "Slide" y up or down

EXAMPLE: "Translate" left 4 and down 3

          *Pick one vertex at a time, "slide" x , then "slide" y

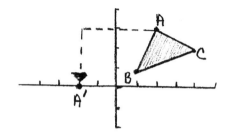

*Do this for all vertices

EXAMPLE:

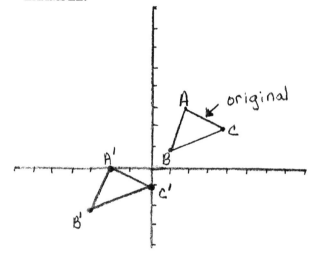

original

original   new transformation

A ⟶ A'

B ⟶ B'

C ⟶ C'

(' means new)

Reflection – must find the ' line of reflection'

*Pick one point at a time, move toward line of reflection and bounce
off the opposite direction the same distance

EXAMPLE: Reflect $y = x$ (line of reflection)

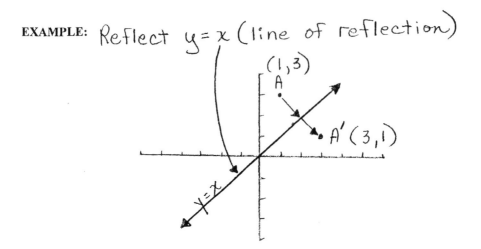

*May be asked to reflect x-axis or y-axis

or

combination reflections
(after a reflection, reflect that reflection)

EXAMPLE:

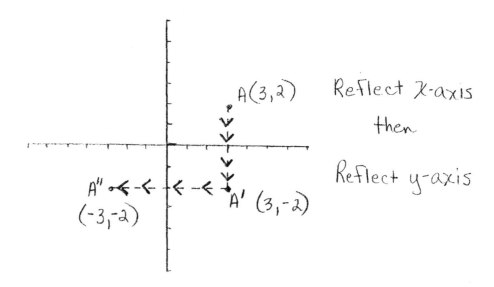

Rotation: easiest to rotate one point at a time, one quadrant (90 degrees) at a time
(pick points on axis if possible)

Clockwise   Counter- Clockwise

EXAMPLE:

Rotate △ABC 270° Counter- Clockwise

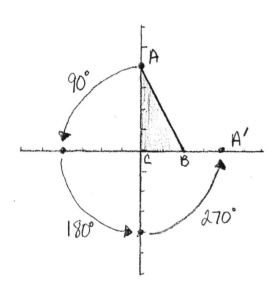

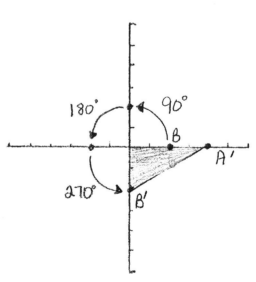

Coordinate Reflections

If (A, B) is reflected across x - axis ⟶ reflection = (A, -B)

If (A, B) is reflected across y - axis ⟶ reflection = (-A, B)

If (A, B) is reflected across y = x ⟶ reflection = (B, A)

If (A, B) is reflected across y = -x ⟶ reflection = (-B, -A)

Coordinate Rotations
About the Origin

*When point (A, B) is rotated clockwise and counter-clockwise about the origin

Clockwise:

(A, B)   Rotated 90 degrees   (A, B) ⟶ (B, -A)

                180 degrees   (A, B) ⟶ (-A, -B)

                270 degrees   (A, B) ⟶ (-B, A)

Counter-Clockwise:

(A, B)   Rotated 90 degrees   (A, B) ⟶ (-B, A)

                180 degrees   (A, B) ⟶ (-A, -B)

                270 degrees   (A, B) ⟶ (B, -A)

Midpoint Formula

*The point in the middle of 2 points on the line – halfway

Formula:
$$\left(\frac{x_1 + x_2}{2} \; , \; \frac{y_1 + y_2}{2} \right)$$

add both
x values;
then, divide by 2
(average x-values)

add both y
values;
then, divide by 2
(average y-values)

EXAMPLE: Find midpoint of AB

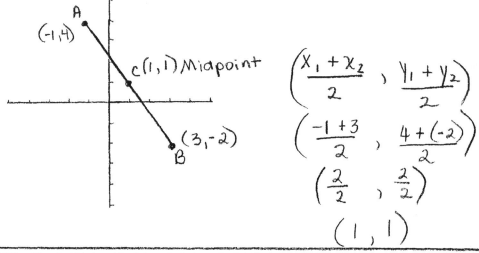

A (-1, 4)

c (1, 1) Midpoint

(3, -2)
B

$$\left(\frac{x_1 + x_2}{2} \; , \; \frac{y_1 + y_2}{2} \right)$$

$$\left(\frac{-1 + 3}{2} \; , \; \frac{4 + (-2)}{2} \right)$$

$$\left(\frac{2}{2} \; , \; \frac{2}{2} \right)$$

$$(1 \; , \; 1)$$

EXAMPLE: If point M is midpoint and A is endpoint, find endpoint B

(1, 1)   (3, 4)

A (3, 4)
endpoint

M (1, 1) midpoint

B
(-1, -2)
missing endpoint

If midpoint is
given, find path
from original
point to midpoint
(left 2, down 3)
Then repeat path to
find missing endpoint.

45

Distance Formula
(Also called "Length")

Distance Formula $d = \sqrt{(x_2 - x_1)^2 + (y_2 - y_1)^2}$

EXAMPLE: Using formula   Find length of AB

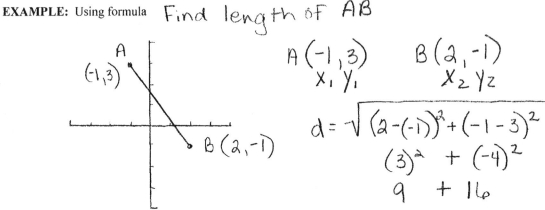

$$A \, (-1, 3) \qquad B \, (2, -1)$$
$$ x_1 \, y_1 \qquad\qquad x_2 \, y_2$$

$$d = \sqrt{(2-(-1))^2 + (-1-3)^2}$$
$$(3)^2 + (-4)^2$$
$$9 + 16$$
$$\sqrt{25}$$
$$d = \boxed{5}$$

Shortcut:

1. Create a right triangle

2. The hypotenuse is always the length or distance asked for

3. Draw a horizontal and vertical line through two vertices to create a right triangle

4. Then use Pythagorean theorem to find C, missing length

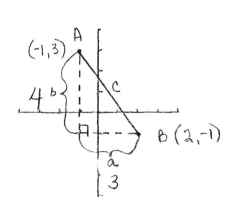

$$a^2 + b^2 = c^2$$
$$3^2 + 4^2 = c^2$$
$$9 + 16 = c^2$$
$$25 = c^2$$
$$\sqrt{25} = \sqrt{c^2}$$
$$\boxed{5 = c}$$

Vectors

Vectors – represented in a coordinate plane by an arrow drawn from one point to another, giving both direction and magnitude

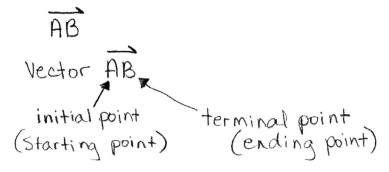

$\overrightarrow{AB}$

Vector $\overrightarrow{AB}$

initial point
(starting point)

terminal point
(ending point)

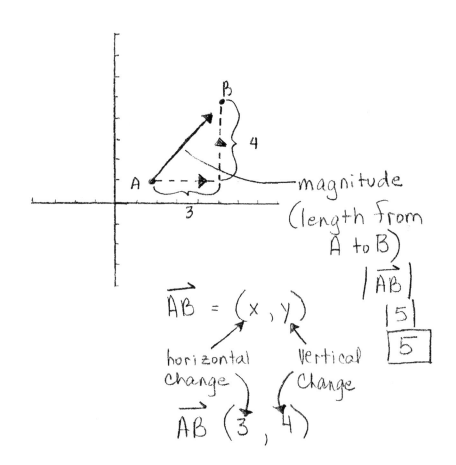

$\overrightarrow{AB} = (x, y)$

horizontal
change

vertical
change

$\overrightarrow{AB}$ (3, 4)

magnitude
(length from
A to B)

$|\overrightarrow{AB}|$

$|5|$

$\boxed{5}$

Circles

WORD PROBLEMS: PIE CHARTS (Always a Proportion)

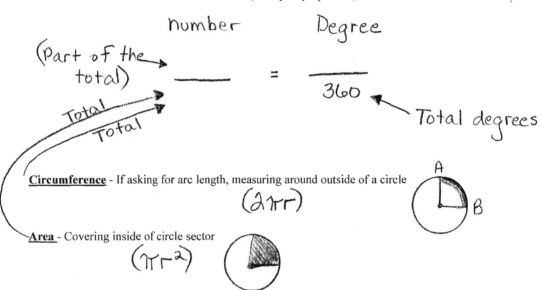

number   percent

(Part of the total) →
(Total) →

$$\frac{\quad}{\quad} = \frac{\quad}{100}$$

↖ Total Percent

*Any of the unknowns can be asked for so read question carefully

WORD PROBLEMS: CIRCLE GRAPHS (Always a proportion)

number   Degree

(Part of the total) →
Total →
Total →

$$\frac{\quad}{\quad} = \frac{\quad}{360}$$

← Total degrees

Circumference - If asking for arc length, measuring around outside of a circle

$(2\pi r)$

A
B

Area - Covering inside of circle sector

(πr^2)

48

EXAMPLE: Find arc length AB (piece of circumference) or find $\overset{\frown}{AB}$

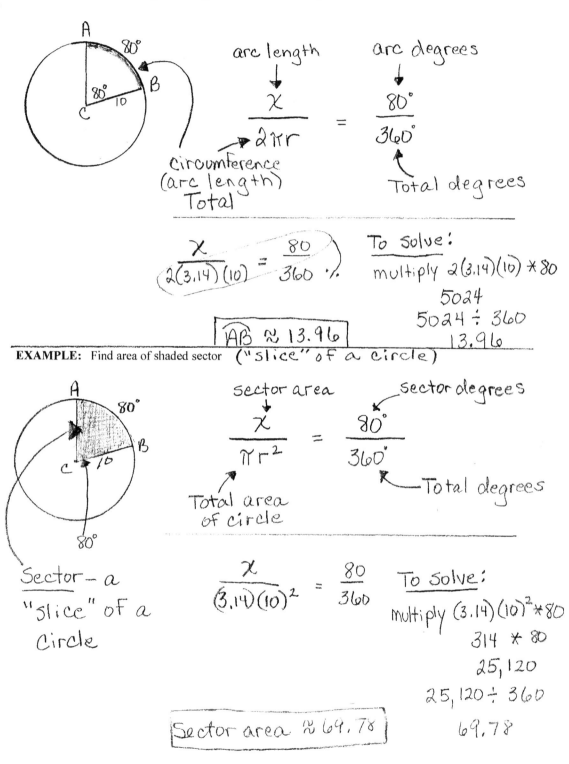

arc length

arc degrees

$$\dfrac{\underset{\downarrow}{x}}{2\pi r} = \dfrac{\underset{\downarrow}{80°}}{360°}$$

circumference
(arc length)
Total

Total degrees

$$\dfrac{x}{2(3.14)(10)} = \dfrac{80}{360} \;\cdot\text{/.}$$

To Solve:
multiply $2(3.14)(10) * 80$

$$5024$$

$$5024 \div 360$$

$$\boxed{\overset{\frown}{AB} \approx 13.96}$$

$$13.96$$

EXAMPLE: Find area of shaded sector ("slice" of a circle)

sector area

sector degrees

$$\dfrac{\underset{\downarrow}{x}}{\pi r^2} = \dfrac{80°}{360°}$$

Total area
of circle

Total degrees

Sector - a
"slice" of a
Circle

$$\dfrac{x}{(3.14)(10)^2} = \dfrac{80}{360}$$

To Solve:
multiply $(3.14)(10)^2 * 80$

$$314 * 80$$

$$25,120$$

$$25,120 \div 360$$

$$\boxed{\text{Sector area} \approx 69.78}$$

$$69.78$$

49

Circle Properties

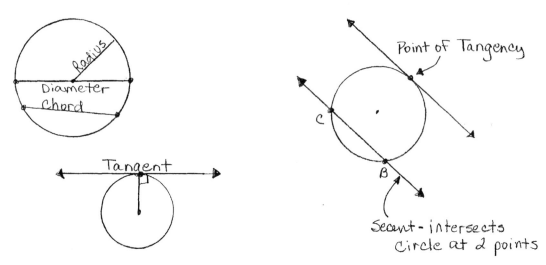

*The intersection of a tangent and radius is **always** perpendicular ($\perp$) – form right angles

---

*Tangents often lead to right triangle problems

EXAMPLE: Line AB is tangent to circle C at point A. Find length of BC

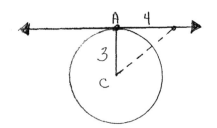

$$a^2 + b^2 = c^2$$
$$3^2 + 4^2 = c^2$$
$$9 + 16 = c^2$$
$$25 = c^2$$
$$\sqrt{25} = \sqrt{c^2}$$
$$5 = c$$
$$\boxed{\overline{BC} = 5}$$

*Tangent segments from an external point are congruent

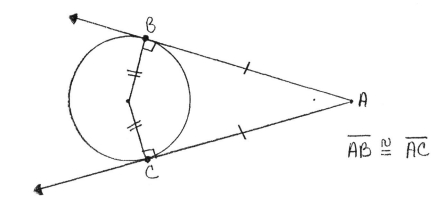

$$\overline{AB} \cong \overline{AC}$$

*Major arcs

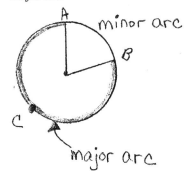

$\overparen{AB} \rightarrow$ minor arc

$\overparen{ACB} \rightarrow$ major arc
(named by 3 points)

*If chord AB is the same distance from center as chord XY, then AB $\cong$ CD   Insert 3

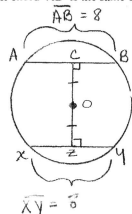

$\overline{AB} = 8$

$\overline{XY} = 8$

IF $\overline{CO} \cong \overline{ZO}$

$3 = 3$

then

$\overline{AB} \cong \overline{XY}$

$8 = 8$

Central Angle – an angle whose vertex is the center of the circle; measure of central angle is same as included arc

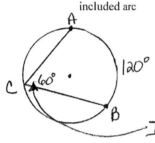

∡ACB is same
as $\overset{\frown}{AB}$
80° = 80°
＊Vise versa

Inscribed Angle – an angle whose vertex is on a circle and whose sides contain chords of the circle; measure of inscribed angle is half the measure of the included arc

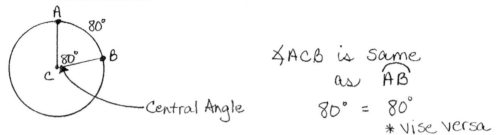

∡ACB is half $\overset{\frown}{AB}$
60° is half 120°

＊Vise versa

EXAMPLE: If $\overset{\frown}{XZ}$ = 80°

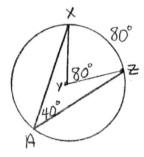

m∡xyz = 80°

m∡XAZ = 40°

＊Vise versa

Chords and Tangents – measure of angle is half the included arc

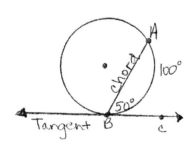

If $\overset{\frown}{AB}$ = 100°
Then
m∡ABC = 50°

＊Vise Versa

52

Secants and Tangents

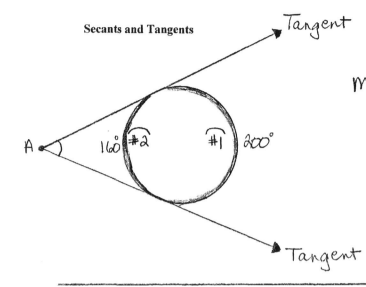

$$m\angle A = \frac{1}{2}\left(\overset{\frown}{\#1} - \overset{\frown}{\#2}\right)$$
$$= \frac{1}{2}\left(200 - 160\right)$$
$$= \frac{1}{2}\left(40\right)$$
$$= 20$$
$$\boxed{m\angle A = 20°}$$

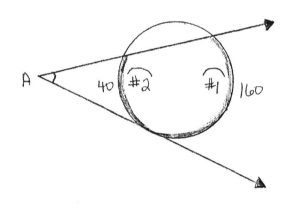

$$m\angle A = \frac{1}{2}\left(\overset{\frown}{\#1} - \overset{\frown}{\#2}\right)$$
$$= \frac{1}{2}\left(160 - 40\right)$$
$$= \frac{1}{2}\left(120\right)$$
$$= 60$$
$$\boxed{m\angle A = 60°}$$

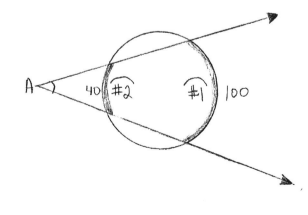

$$m\angle A = \frac{1}{2}\left(\overset{\frown}{\#1} - \overset{\frown}{\#2}\right)$$
$$= \frac{1}{2}\left(100 - 40\right)$$
$$= \frac{1}{2}\left(60\right)$$
$$= 30$$
$$\boxed{m\angle A = 30°}$$

Chords

Lengths

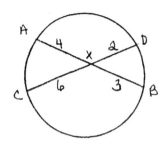

If 2 chords intersect,
$$Ax * Bx = Cx * Dx$$

$\underbrace{}$

Products are
equal
$$2 * 6 = 3 * 4$$
$$12 = 12$$

Arcs and Angles

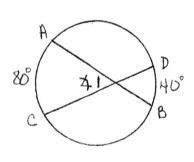

$$m\angle 1 = \tfrac{1}{2}\left(\overarc{AC} + \overarc{DB}\right)$$
$$m\angle 1 = \tfrac{1}{2}\left(80 + 40\right)$$
$$= \tfrac{1}{2}\left(120\right)$$
$$m\angle 1 = 60$$

*If two secant segments share a common endpoint outside a circle, product of the lengths of one secant segment and its' external segment equals the product of the length of the other secant segment and its' external segment

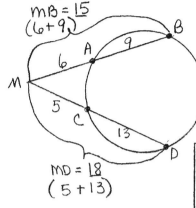

$$MA * MB = MC * MD$$
$$6 * 15 = 5 * 18$$
$$90 = 90$$

EXAMPLE:

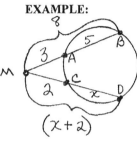

$$MA * MB = MC * MD$$
$$3 * 8 = 2(x + 2)$$
$$24 = 2x + 4$$
$$-4 \qquad\qquad -4$$
$$\frac{20}{2} = \frac{2x}{2} \qquad \boxed{10 = x}$$

*If secant and tangent segments share a common endpoint outside a circle, product of secant and external segment equals square of the lengths of tangent

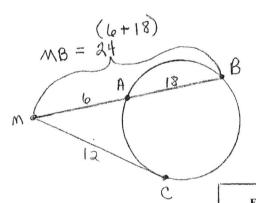

$$MA * MB = (MC)^2$$
$$6 * 18 = 12^2$$
$$144 = 144$$

EXAMPLE:

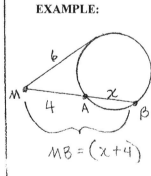

$MB = (x + 4)$

$$MA * MD = (MC)^2$$
$$4(4 + x) = 6^2$$
$$16 + 4x = 36$$
$$-16 \qquad\qquad -16$$
$$\frac{4x}{4} = \frac{20}{4}$$
$$\boxed{x = 5}$$

Circumcenter

*The point of concurrency of the three perpendicular bisectors of a triangle

Acute Triangle – P is inside triangle

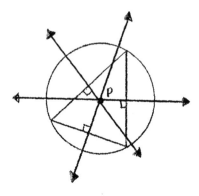

Right Triangle – P is on triangle

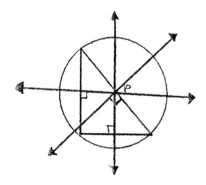

Obtuse Triangle – P is outside triangle

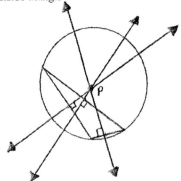

Orthocenter

*A point at which the lines containing the 3 altitudes intersect

Acute Triangles -- Orthocenter is inside triangle

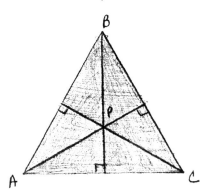

Right Triangles – Orthocenter is on triangle

Obtuse Triangles – Orthocenter is outside triangle

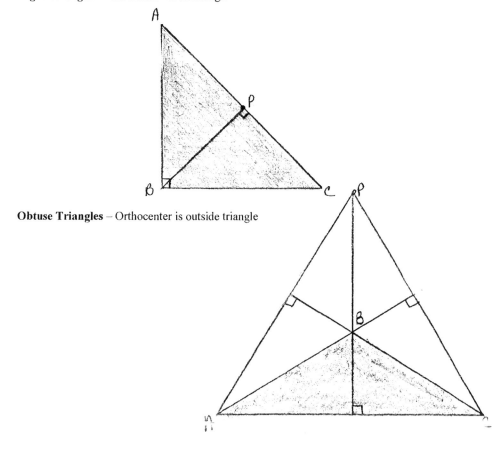

*An altitude of a triangle is the perpendicular segment from a vertex to an opposite side

Acute Triangle

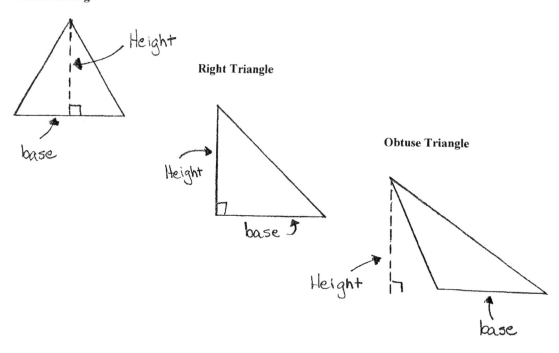

Right Triangle

Obtuse Triangle

Centroid

*A segment from a vertex to the midpoint of the opposite side

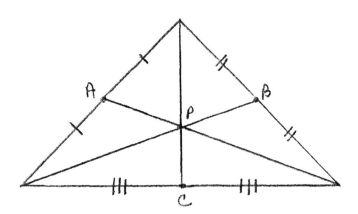

Coplanar Circles

Tangent Circles – Coplanar circles that intersect at one point

 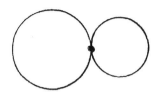

Concentric Circles -- Coplanar circles that have a common center (have no points of
Intersection)

*No points of intersection

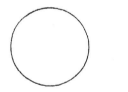

*Two points of intersection

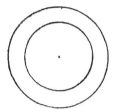

Common Tangents—A line, ray or segment that is tangent to two coplanar circles

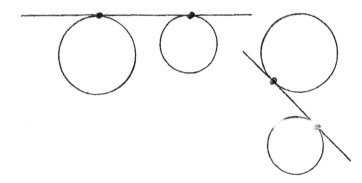 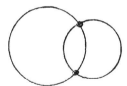 

Areas

Square: Area = square of the side lengths (or product of base and height)

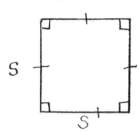

$S = Side$

$A = S^2$

or

$A = S * S$

$A = 5 * 5$

$\boxed{A = 25}$

Rectangle: Area = product of base and height

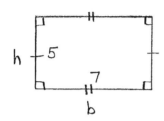

$h = height$
$b = base$

$A = b * h$

$A = 7 * 5$

$\boxed{A = 35}$

Parallelogram: Area = product of base and height

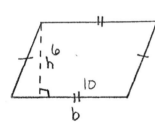

base and height <u>always</u> form a right angle

or

are perpendicular

$A = b * h$

$A = 10 * 6$

$\boxed{A = 60}$

Triangle: Area = ½ product of base and height

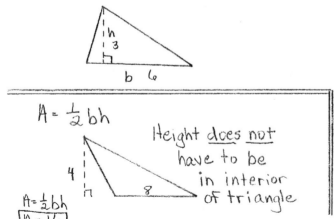

$A = \frac{1}{2} bh$

$A = \frac{1}{2} * 6 * 3$

$A = \frac{1}{2} * 18$

$\boxed{A = 9}$

$A = \frac{1}{2} bh$

Height does <u>not</u> have to be in interior of triangle

$A = \frac{1}{2} bh$

$\boxed{A = 16}$

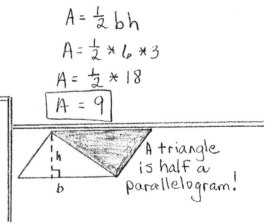

A triangle is half a parallelogram!

60

Trapezoid: Area = ½ product of the height and sum of the lengths of the bases

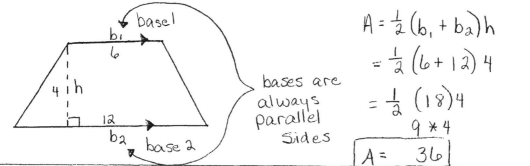

$$A = \frac{1}{2}(b_1 + b_2)h$$
$$= \frac{1}{2}(6 + 12)4$$
$$= \frac{1}{2}(18)4$$
$$9 * 4$$
$$A = 36$$

Rhombus: Area = ½ product of length of diagonals

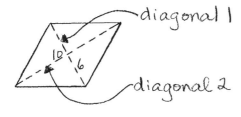

$$A = \frac{1}{2}(d_1 * d_2)$$
$$= \frac{1}{2}(6 * 10)$$
$$= \frac{1}{2}(60)$$
$$A = 30$$

Kite: Area = ½ product of the lengths of diagonals

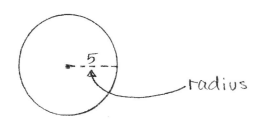

$$A = \frac{1}{2}(d_1 * d_2)$$
$$= \frac{1}{2}(4 * 12)$$
$$= \frac{1}{2}(48)$$
$$A = 24$$

Circle: Area = ___ times the square of the radius

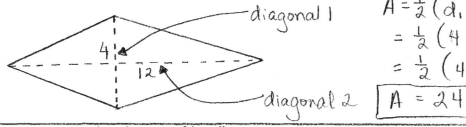

$$A = \pi r^2$$
$$= (3.14)(5)^2$$
$$= (3.14)(25)$$
$$A = 78.5 \text{ or } 25\pi$$

Area of Regular Polygon: Area = ½ product of apothem and perimeter

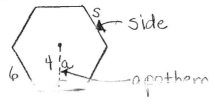

6 sides

$$A = \frac{1}{2}aP \quad \text{(number of sides)}$$
$$= \frac{1}{2}(4)(6 * 6)$$
$$= \frac{1}{2}(4)(36)$$
$$= \frac{1}{2}(144$$
$$A = 72$$

Solids
Volume and Surface Area

Prisms: Bases are congruent polygons lying in parallel planes
The other faces are lateral faces and are parallelograms

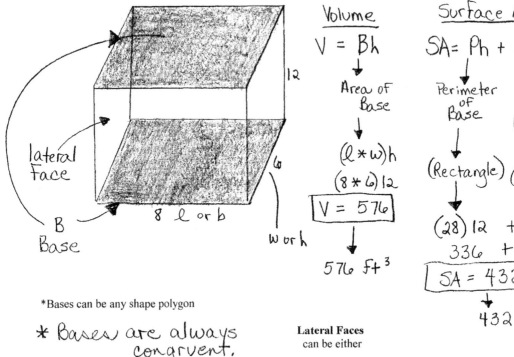

12

6

8 ℓ or b

lateral
Face

B
Base

w or h

Volume

$$V = Bh$$

↓

Area of
Base

↓

$(\ell * w)h$

$(8 * 6)12$

$\boxed{V = 576}$

↓

576 ft^3

Surface Area

$$SA = Ph + 2B$$

↓ Perimeter of Base   ↓ Area of Base

(Rectangle) (Rectangle)

↓

$(28)12 \quad + \quad 2(48)$

$336 \quad + \quad 96$

$\boxed{SA = 432}$

↓

432 ft^2

*Bases can be any shape polygon

* Bases are always
      congruent.
* Bases can be any shape polygon.

Lateral Faces
can be either

Right
Lateral edges are
are perpendicular
to bases

Oblique
Lateral edges are
are **NOT** perpendicular
to bases

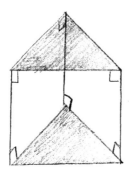

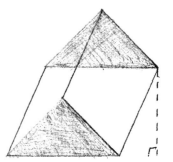

Regular Pyramids: Square base and 4 congruent faces with a common vertex

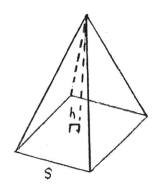

$$V = \frac{1}{3} Bh$$

Example: $s = 6 \quad h = 10$

$$V = \frac{1}{3} Bh$$
$$\downarrow$$
$$\frac{1}{3} \cdot s^2 \cdot h$$
$$\frac{1}{3} (6)^2 (10)$$
$$\frac{1}{3} \cdot 36 \cdot 10$$
$$12 \cdot 10$$
$$120 \text{ units}^3$$

Cones: circular base and a vertex that is not in the same plane as the base

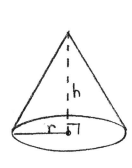

$$V = \frac{1}{3} Bh$$

Example: if $r = 3 \quad h = 8$

$$V = \frac{1}{3} Bh$$
$$\downarrow$$
$$\frac{1}{3} \cdot \pi r^2 \cdot h$$
$$\frac{1}{3} \pi (3)^2 (8)$$
$$\frac{1}{3} \cdot \pi \cdot 9 \cdot 8$$
$$24\pi \text{ units}^3$$

Cylinders: Solid with congruent circular bases that lie in parallel planes

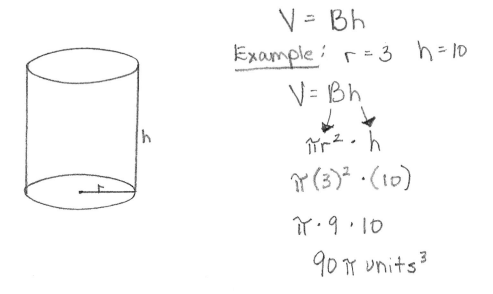

$$V = Bh$$

Example: $r = 3$ $h = 10$

$$V = Bh$$

$$\pi r^2 \cdot h$$

$$\pi (3)^2 \cdot (10)$$

$$\pi \cdot 9 \cdot 10$$

$$90\pi \text{ units}^3$$

Spheres: A set of all points in space that are equidistance from a given point

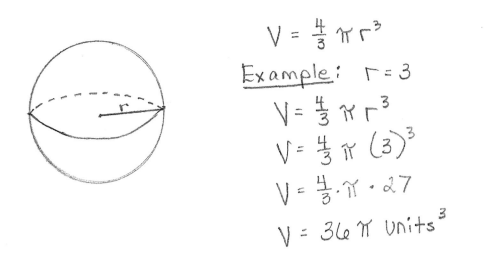

$$V = \frac{4}{3}\pi r^3$$

Example: $r = 3$

$$V = \frac{4}{3}\pi r^3$$

$$V = \frac{4}{3}\pi (3)^3$$

$$V = \frac{4}{3}\cdot \pi \cdot 27$$

$$V = 36\pi \text{ units}^3$$

About the Author

I have taught math from the 6th grade level through college for over 30 years. I am presently teaching day and night at an alternative high school in San Marcos, Texas, I have my own private tutoring business, and teach ACT and SAT prep classes on Saturdays. I know that math can be very overwhelming and just plain scarey! My favorite saying to students is, "How do you eat an elephant? One bite at a time, of course!" That 'one bite' is a solid definition or rule, followed by a simple example. From this, all math problems can be mastered. I also include many shortcuts, tricks and silly ways to remember skills easily and quickly. If you take "one bite at a time," before you know it, the elephant will be gone!